Congratulations, by the way

人生最好的毕业礼

George Saunders

〔美〕乔治·桑德斯 著

徐之野 译

南海出版公司

目录

献词
01

作者简介
02

出版说明
04

话题热议
06

名校毕业演讲
16

导读：迈入社会之前，最重要的事是什么
28

人生最好的毕业礼
39

乔治·桑德斯写给年轻人的话
95

媒体和名人推荐
116

读者推荐
122

附　写给毕业五年后的自己
129

英文版原文
132

作者简介

乔治·桑德斯
George Saunders

美国当代杰出短篇小说家,《纽约时报》畅销作家,美国雪城大学艺术与科学学院英语教授,《纽约客》《GQ 杂志》《哈泼氏杂志》专栏作家。其小说《林肯在中阴界》获得 2017 年布克奖;《十二月十日》荣获第一届弗里欧文学奖,并成为《纽约时报》畅销书。

乔治·桑德斯被《时代》杂志(Time)评选为 2013 年度全球百大影响力人物之一。起先,桑德斯以畅销童书《小魔怪黏巴达》(The Very Persistent Gappers of Frip, 2000)获得图书奖项肯定。后来他致力于短篇小说的创作,并在国际文坛引起瞩目,四度获得美国国家杂志奖,并先后获得欧·亨利小说奖、杰出短篇小说奖以及美国文化界最高奖项——麦克阿瑟天才奖,被称为"作家中的作家"。

桑德斯有许多作品延续了乔治·奥威尔式的荒谬讽刺风格，因为曾经在柯达及弧度（Radian，环境工程公司）等大公司工作的亲身体验，他的作品对资本主义追名逐利、冷漠无情的现状多有批判，但又能兼具诙谐幽默。他的小说也经常被拿来跟美国黑色幽默文学的代表作家库尔特·冯内古特相比，相较于冯内古特的愤世嫉俗，桑德斯的作品则对世界充满了希望。他曾通过笔下一位失意沮丧的人物说：

"继续活下去、与亲人保持联系，因为'生活中依然可能有很多点点滴滴的善良'。"

出版说明

每当毕业季开始,美国各大学府就会邀请重要的社会人士莅临,为将要踏上梦想之路的年轻人进行演讲。这些来自各行各业的演讲者常常会提到哪些经历影响了他们的人生道路,哪些想法支持他们走到现在。例如,苹果集团创始人乔布斯在斯坦福大学说过"追求卓越永无止境,勿忌别人品头论足";亚马逊董事会主席贝佐斯在普林斯顿大学以一段年轻时的往事劝说世人:对很多人来说,同理心比聪明更困难……而本书则是美国著名小说家乔治·桑德斯在母校雪城大学(Syracuse University)的演讲内容。

演讲中,桑德斯谈起他人生中最后悔的一件事。即使犯过年轻人都会犯的错、闯过年轻人都会闯的祸,历经岁月沉淀后,一件往事还是令他念念不忘。他反思自己——为什么没能及时回应他人的需要,为什么不能及时做该做的事情……演讲发表后三个月,他的演讲全文由《纽约时报》

刊登到网站上,一天之内,即被深受感动的百万读者分享传播!

为什么这则简单的小故事会引起这样大的震动?因为这些话其实就是许多人内心对生活的渴望:渴望拥有勇气,渴望人与人之间有更多善意,期待与人真诚交流可以带来生命的意义。

如今我们将其编辑成中英文对照版,将这一隽永的讲稿献给即将踏入社会的年轻人,提醒大家:这一生除了时时让自己对世界怀抱善良之念,更要勇敢追梦。

希望这份特别的毕业礼对世人有更珍贵的价值。恭喜毕业,恭喜!

话题热议

💡 许多人大学毕业时也许只想听到接下来该怎么赚钱的话,但如果要在赚钱之道之外听到一点真诚坦率的人生箴言,那没有比这本书做得更棒的了。

——**《娱乐周刊》**

💡 桑德斯对人们建议去无私地爱和他所指出的人与人之间的相互关系,实在是最纯净、最简单,但也是最艰巨的人生道理。

——**《科克斯评论》**

💡 暖心又温柔的一本书。

——**《出版人周刊》**

💡 我们比想象中的强大。我们在学会拥有勇气之前并不知道自己有多么勇敢。我们可以战胜一切困难,甚至那些自己没遇到过的困难。是什么让我们能够利用这些力量和勇气,发挥自己的最大潜能?是全情投入,勇敢和塑造自我生活的决心。

——投资家、伯克希尔 – 哈撒韦公司首席执行官
沃伦·巴菲特

💡 选择希望而不是恐惧,这是需要勇气的。人们或许总会说你幼稚,但是我们向前迈出的每一步,都离不开这份希望和乐观。

——Facebook 创始人兼首席执行官　马克·扎克伯格

- 我觉得最大的经验就是千万不要放弃,要勇往直前,而且不断地创新和突破;突破自己,直到找到一个方向为止,而且我觉得还有更重要的一点:我们今天面对将来的信心是来自我们前五年的残酷经验,我们坚信明天更加残酷。

   —— **阿里巴巴集团创始人　马云**

- 我们要有勇气去改变可以改变的事情,有胸怀去接受不能改变的事情,并且有智慧来分辨这两者。

   —— **创新工场董事长兼首席执行官　李开复**

- 我们的生命需要什么呢?突破,突破,再突破!有时,挡住我们前进的脚步,恰恰是不愿意迈出第一步的自己。……成功不在于坚持了多久,只有在一次一次没有希望的时候依然坚持下去,才有用。

   —— **新东方教育集团董事长、中国青年企业家协会副会长　俞敏洪**

做引领者是时代赋予你们的使命。引领是一种境界，境界越高，格局越大。"不畏浮云遮望眼，只缘身在最高层"。只有站在高处，才有不同于别人的视野，才能看见别人看不见的远方。引领也是一种担当，需要十足的勇气，有敢为天下先的气魄，敢于发出自己的声音，敢于迈出第一步。做引领者，要承担比常人更大的压力，要冒更大的风险。

——清华大学前校长　邱勇

要坚守承诺、执着追求，要敬畏自然、担当责任。一辈子很长也很短，眨眼之间就是三十年、四十年、半个世纪，人生的精彩，也就在这执着的选择和坚守之间。

——北京大学校长　林建华

- 假如你拥有高尚的情操,过着俭朴的生活,并且存谦卑的心,那么你的生活必会非常充实。你会是个爱家庭、重朋友,而且是关心自己健康的人。你不会着意社会能给你什么,但会十分重视你能为社会出什么力。

    —— **香港中文大学前校长　沈祖尧**

- 学校毕业,正是生涯旅程的开始,且为行囊中添加智慧。

    —— **台湾清华大学校长　贺陈弘**

- 诚朴以立,止于至善,做一个体贴、感恩、慈悲的世界公民。

    —— **台湾中央大学校长　周景扬**

💡 第一,希望你们扎根自己的领域王国,像蜗牛一样"一步一步往上爬",不要让"我的青春我做主"成为频繁跳槽的借口;第二,希望你们勇于挑战极限,不要做"差不多先生"和"还凑合小姐";第三,希望你们做顶天立地的"小巨人",不要只想着自己的"小确幸"和"小时代";第四,希望你们"人生的巨轮"永远不沉,不要做玻璃心的"小公主",让"人生的小船"说翻就翻。最后,愿你们每一个人都能做一位出彩的人生"工匠大师"!

——**教育部副部长、武汉大学原校长　李晓红**

💡 世事虽复杂,但有"容"则"易"。多一些包容,少一些争斗,我们的工作就会轻松些,交往就会简单些,身心就会健康些。有多大的包容,就有多高的境界;有多高的境界,就能干多大的事业。

——**南京大学校长　陈骏**

💡 一定要在人生的内存，给自己、给至爱的人，留一个百分之一的空间，不随波逐流，哪怕是一个爱称。……要做到这样的百分之一，我的建议是永远带走和大学的脐带关系。这种脐带关系，最重要的是一种精神，那就是永远的批判精神。不迷信，不盲从，不崇拜任何东西。永远对现状不满足，永远想改造世界，也永远拥抱世界上的美好——因为大学培养的是二三十年后国家和人类的领导者和创造者！当你和大学保持这样的脐带关系，你到50岁后，还会激荡青春的豪情；就是到80岁，还有一个不老的灵魂。

——厦门大学教授　邹振东

名校毕业演讲

旁观不是你想要的生活,世界需要你登上舞台,有很多问题需要解决,正义需要得到伸张,人们依然受到迫害,疾病依然需要治愈。无论你接下来要做什么,世界需要你付出能量、激情和成功的渴望。不要怕冒险,远离那些愤世嫉俗者和批评者。历史很少由一个人来书写,但永远不要忘记,历史的确曾由一个人来书写。

——苹果公司现任首席执行官 蒂姆·库克

乔治·华盛顿大学毕业礼演讲

你必须要相信某些东西：你的勇气、目的、生命、因缘……这个过程从来没有令我失望，只是让我的生命更加地与众不同。……你们的时间很有限，所以不要将它们浪费在重复其他人的生活上。不要被教条束缚，那意味着你和其他人思考的结果一起生活。不要被其他人喧嚣的观点掩盖你真正的内心的声音。还有最重要的是，你要有勇气去听从你直觉和心灵的指示——它们在某种程度上知道你想要成为什么样子，所有其他的事情都是次要的。

—— **苹果公司联合创始人　史蒂夫·乔布斯　斯坦福大学毕业礼演讲**

你们都面对着一个充满不确定性的世界。不要害怕这种不确定性,去拥抱它,去利用它。不确定性意味着没有什么是预先确定的。不确定性意味着未来需要你们去塑造——用你们的力量,你们的意志,你们智慧的力量,你们同情心的力量。不确定性就是自由,抓住这种自由,带着它奔跑吧,别忘了一边跑一边吃点面条,你需要葡萄糖。

——世界银行行长 金墉 美国东北大学毕业礼演讲

事实、证据、原因、逻辑以及对科学的理解,这些才是好品质。这些才是政策制定者所应该拥有的品质,是你们把自己培养成优秀公民所需要的品质。……对事实、理性和科学的排斥,会导致退步。这让我想起了卡尔·萨根(美国天文科普作家)讲过的话,他高中毕业于这里——新泽西,他说:判断我们是否进步,要看我们是否有提问的勇气以及解答问题的深度,还有我们对真相的接纳,而非那些让人感觉良好的东西。……当然,我并不是说冰冷的分析和数据,比生活中的激情、信念、爱和忠诚更重要。我想说的是,这些高级的人性,只有在经济运行良好、预算增加、环境变好时才能畅快淋漓地展现。

——美国前总统　奥巴马　罗格斯大学毕业礼演讲

你无法事前画好前进的道路,任何前进的道路都不是事先画好的。你需要弄清楚,你爱做什么,对什么有信念,你就做什么。这很有挑战性,到目前为止,你们这些人,都是通过达到和超越期望来获得今天的成就。你们很棒,你们很优秀。看看你们,你们就像一个惊人的巨型合唱团,但从此以后,你们需要转变角色,你们将不再只是达到和超越期望。没有期望,没有剧本。做你所爱的事情时,你会变得富有塑性,因为这就是你为自己养成的习惯:你养成了勇于冒险的习惯,大胆选择你所爱做之事的习惯。

——Twitter 前首席执行官 迪克·科斯特洛
密歇根大学毕业礼演讲

一、想法很重要。领导、组织和辛勤工作都很重要,但想法决定一切。二、去生活的健身房,锻炼新能力,让自己更快更好。三、学会如何讲故事:用一种简练明晰的方式,将你心中想法的价值传递给他人。讲故事是经营中最被低估的技能之一。四、珍视时间。作为世界未来的领袖,你们需要不断询问自己,什么样的时间框架是正确的,应该怎样考虑各种时间的投资。五、做人要"硬核":要有激情,要坚毅,要乐观,要有鲨鱼特质(要么往前,要么死)。六、追求快乐。

——微软公司前首席执行官兼总裁 史蒂夫·鲍尔默
南加利福尼亚大学毕业礼演讲

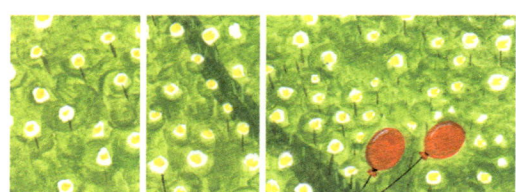

今天我要告诉你们的是"天赋"和"选择"之间的差别。聪明是一种天赋,而善良是一种选择。天赋得来容易,但选择往往很困难。……我要斗胆做个预测。当你们活到 80 岁,在某个安静的沉思时刻,回到内心深处,想起自己的人生故事时,最有意义的部分,将会是你所做过的那些选择。人生到头来,我们的选择,决定了我们是什么样的人。

——**亚马逊集团董事会主席兼首席执行官　杰夫·贝佐斯**
普林斯顿大学毕业典礼演讲

记住,人生的设计目的不是为了折磨你,而是旨在给你快乐和享受。你应该寻求享受生活。只有在不时的磨难和痛苦时还在享受人生的人,才会留下可以持久的印记,拥有充实和有意义的人生。

——**凯雷投资集团联合创始人　大卫·鲁宾斯坦**
杜克大学毕业礼演讲

如果你们选择用你们的地位和影响力来为没法发出声音的人说话；如果你们选择不仅认同有权的强势群体，也认同无权的弱势群体；如果你们保留你们的能力，用来想象那些没有你们这些优势的人的现实生活，那么不仅是你们的家庭为你们的存在而感到自豪，为你庆祝，而且那些因为你们的帮助而生活得更好的数以千万计的人，会一起来为你们祝贺。我们不需要魔法来改变世界，我们已经在我们的内心拥有了足够的力量：那就是把世界想象成更好的力量。

——《哈利·波特》作者　J.K. 罗琳　哈佛大学毕业礼演讲

希望你们能尽一己之力，使我们免于恐惧。……从现在开始，你们终身都将背负着一份责任，这是你们的终身使命，是你们身为人类、美国人和耶鲁毕业生的职责，就是站在恐惧和信念的杠杆支点上。恐惧在你后方，信念在你前方。你会往哪个方向倾斜？你会往哪个方向移动？请勇往直前，一路勇往直前，然后用 Twitter 传播你努力的成果。搞不好你会像塞缪尔（美国 NBC 主持人，人称"耶鲁哥"）一样名满天下。

——**美国影星　汤姆·汉克斯　耶鲁大学毕业典礼演讲**

最重要的是,社会会倾听你们的呼声。如果你们中的一员遭遇不幸,尤其是如果涉及可能的不公对待的话,这会在一夜之间成为重大新闻。这是你最重要的特权:作为社会上最具良好教育的群体,你们的言行、你们的世界观价值观,很快就会成为社会主流思潮。如果你们希望社会进步,请按照你期待向往的价值观生活,而不是去迎合固化现有的思潮。

——哥伦比亚大学商学院副院长　姜纬　复旦大学毕业礼演讲

我们需要储备多样化的技能和知识,而更重要的是,要有一颗开放和包容的心去学习,这就是为什么我们非常幸运地成为凯斯特罗姆商学院(波士顿大学的下属商学院)群体的一部分,因为它培养了开放思想的自由。

——波士顿大学 2017 年中国毕业生　蔡语婧
　　波士顿大学毕业礼演讲

导读：迈入社会之前，最重要的事是什么

如果问朝气蓬勃、对未来抱有无限憧憬的准毕业生们一个问题：迈入社会之前，最重要的事是什么？相信无数的答案会汹涌而至：找个像样的工作；拿到可以证明能力的若干资格证；学习职场技能、沟通能力；梳理学校人脉关系网；和往事干杯甚或抓紧时间和妹子表白……

如果说社会是复杂的，那学校大门隔开的是截然不同的阶段：理想与现实，单纯与复杂，自我与责任，青春与成熟……但无论处在人生的哪个阶段，大部分道理总是相通的：工作和生活需要方法，也需要努力；Offer、学历、平台、机遇、背景之类的硬件不容忽视，沟通力、执行力、领导力等软件也很重要……而灵魂的开关机方式是一切的重中之重。我们用什么形式来为人生开关机？读下去，相信会给你带来一份特别的启迪。

有些书篇幅不长，给予世人的震撼却不小，《人生最好的毕业礼》就是这样的作品。作者乔治·桑德斯入选《时代》杂志 2013 年度全球百位最具影响人物，被誉为"用英文写作的最优秀的短篇小说家"，是布克奖、欧·亨利奖和麦克阿瑟奖（俗称"天才奖"）得主……尽管这位天才人物获奖无数，却都不及他在雪城大学那次 10 分钟的演讲带给人们的震撼大。那天，乔治·桑德斯回忆了一件小事：

他上七年级时，班里有一个矮矮的、羞涩的小女孩，总戴着老式眼镜，常常因紧张而下意识地把一绺头发咬进嘴里，原本被同学们忽视的她因此得到了一点"关注"——嘲笑。女孩孤独地煎熬着每一天，直到转校，只是旁观的那个沉默怯懦的小男孩乔治·桑德斯多年以后仍不能忘怀。

尽管当时真的"没有发生什么大悲剧",而这在桑德斯看来,正是心结所在——面对欺凌和冷暴力,明知对方那么无辜,那么无助,却始终待在自己的安全区,以致随波逐流。或因为怯懦,或因为恐惧,他始终没有伸出援助之手,哪怕自己心里并不好受。其实,只要说一句话或许就可以引开那些调皮男孩的注意力,帮助女孩走出窘迫,他却没有去做……

有些话看似简单,但缺乏勇气便永远说不出口;有些事看起来与己无关,但袖手旁观却会让自己身心不安。勇气有时候是一辈子的执念,有时候是一瞬间的闪念。人生,无非是矛盾与选择的过程,无关对错,有时候仅仅在于面对矛盾或纠葛时我们能否有勇气进行选择,并勇敢承担后果。

我们不妨在记忆之河中检索一番，是否有类似的不敢触碰的疮疤？那一刻的胆怯与退缩沤在心里许久，似乎已经忘却，却断续溢出焦灼。恐惧和胆怯犹如扎在心尖上的刺，不能幻想它自己会被顶出来，而是得主动拔掉它。紧接着的问题，是知道这一切之后你会怎样？是继续独善其身，并在充满变数的将来变得麻木而平庸，还是在踏入社会之前立刻做出改变，战胜恐惧、自卑等让人失落悲观的东西？本书告诉我们：于我们自身，唯有以勇气挑战自己，才会超越自我，实现内在自得；于他人和世界，唯有心怀善意，才能摆脱所谓"成功"的束缚，成就真正的自我；于人际、职场，唯有以勇气挑战障碍和不公平，才会不断精进。简单来说，就是：勇敢做自己，善良对他人。就像乔治·桑德斯说的那样："为了成为最出色的自己，我们必须怀着雄心，把自己最好的一面活出来。"

生活本身是沉重的，听到乔治·桑德斯"如诗篇一样轻盈，但却深厚无与伦比"（《纽约时报》）的一席智言却是幸运的，从北美到全球，社交网络将雪城大学毕业仪式中那10分钟的"惊雷"，在最初聆听演讲的千百名学子间绵延传递，继而爆炸式地在数亿人中引发强烈反响。由该演讲集结成书的《人生最好的毕业礼》自出版以来，温暖了无数迷茫的年轻人，书中那个简单而振奋人心的小故事，除勇气和善良之外，还涉及了成功、爱，甚至世间万物的道理，是一份送给年轻人的人生礼物。

无论是不满还是不舍，是快乐还是遗憾，在学校的日子犹如溪流般清澈透明，稍纵即逝。如今，我们即将告别这段美好而隽永的历程，迎向未知的茫茫前路。路途中，可能会阳光明媚，但也可能风沙漫天。对即将毕业的年轻人来说，拥有善良和勇气至关重要，因为这充满挑战的世界更喜欢善良且有勇气的年轻人。斯蒂芬·茨威格说："勇气如逆境中绽放的光芒，它是一笔财富，拥有了勇气，就拥有了改变的机会。"在善念之上拥有勇气，才可以面对前路各种未知和内心的纠缠；而基于勇气之上坚持善良，我们所做的坚持才能定义为信念，我们的灵魂才会装上羽翼，未来才可以拥有无数更好的可能。

年轻人，请接受这份毕业礼——用一颗充满勇气的灵魂开启你无限的未来！

人生最好的毕业礼

——乔治·桑德斯雪城大学演讲

随着时代变迁，给大学毕业生做演讲慢慢发展出一种固定模式：邀请某个风光不再，在人生中犯过各种大错的傻老头（就像我），给一群风华正茂、精力旺盛、前途无限的年轻人（就像你们）一点诚恳实在的建议。

今天我的发言准备遵照这个模式。

现今，老一辈对年轻人的用处，除了向他借钱花，或是请他表演上个时代的舞步，好让台下的你们观赏取乐之外，你还可以问他："回首过去大半生，你有什么觉得后悔的？"这个时候，他会有好多话告诉你。有时候，你知道的，就算你不问，他也会抢着说。甚至你要求他别说，他还是停不住。

那么，我后悔什么呢？

后悔我经常把自己搞得太穷?
不是。

后悔我干过一些糟糕透顶的工作——比方说跑到屠宰场里给生肉去骨(我完全不想谈这个工作的细节)?
也不是。

我并不后悔干过这些事。

我曾经在苏门答腊岛的河流中裸泳。当时我听到一阵吵闹声，抬头一看，有三百多只猴子正在河道旁往河里大便，我惊讶得张大嘴巴，一丝不挂地愣在河里，还因此染上重病，整整七个月才痊愈。我后悔当时下水吗？现在想想，我不怎么后悔这档事。我后悔做过一些让自己蒙羞的事吗？比如我在一大群观众前打曲棍球，当时我很喜欢的女孩也坐在观众席中，一不留神我摔倒了，我哇哇怪叫的同时，鬼使神差地把那颗球打进了自己队的球门，并且球棒被我打飞，奔向观众，险些砸中那个女孩——不，我一点也不会把这件事放在心上。

真正让我后悔莫及的是这一件事：

"那一年我上七年级，班里转来了一个新同学。为了保护对方的隐私，我在这里暂时给她取个化名叫艾伦。艾伦个子很小，个性羞涩。她戴着蓝色像猫眼的眼镜——那时候只有老太太才会戴这种眼镜。她很容易紧张，一紧张就不由自主地把一绺头发含在嘴里嚼来嚼去。"

总之，这个艾伦来到我们学校，住进我们的社区，不过人们经常忽略她，还时常嘲笑她（他们会说"你的头发很好吃吧？"这一类的话）。我知道这些事情对她造成了伤害。我还能记得她被嘲笑侮辱后脸上的神情：低头望向地面，似乎有点自责，好像被人提醒了她的处境，拼命想立刻在人们面前消失。她被嘲弄之后，就会走开，嘴里依然含着那一绺头发。我在自己家里会想，如果她放学回家，她妈妈像别人母亲一样问起："宝贝，今天在学校开心吗？"她大概会回答："嗯，很好。"她妈妈会再问她："今天交到了朋友吗？"她则会回答："对，很多朋友。"

有时候，我看到她独自一人在自家前院里踱步，似乎害怕离开那个院子。

……后来她们就搬家了。就这样。没有发生什么大悲剧，也没有重大灾难。

前一天她还在学校里，第二天她就不在了。

故事结束了。

现在你们要问，我有什么好后悔的呢？为什么四十二年之后，我还对这件事耿耿于怀？相比学校里大多数人，我对她算得上友善。我从未对她出言不逊，甚至有几次我还挺身为她辩护过几句。

然而，我就是至今对这件事耿耿于怀。

因此，我发现一个真实的道理——虽然听起来有点老生常谈，我不知道还有什么高明的说法：

我一生中最后悔的事就是没有及时表达对他人的善意。

当一个遭受痛苦的人在我面前，当我看到这样的一幕，我的回应却是……理性克制、沉默寡言，甚至无动于衷。

又或者，从事情的另一方面来看：谁，在你一生的记忆里，会让你深深爱惜惦念着；只要想到他，就会毋庸置疑地感到温暖？

我相信是那些对待你最仁慈善良的人。

与人为善看似轻而易举，实行起来却困难重重。我觉得，如果把这作为人生的目标，不管怎么努力，我们常常连"善良一点儿"都做不到。

所以，最重要的问题来了：

问题出在哪里？

为什么我们总是不够善良呢？

以下是我的看法：

打从出生，我们的脑海中不免带着一连串与生俱来的错误观点，这大概在某种程度上符合达尔文的适者生存法则，带着谬误的人才能存活至今——我们搞错的事情不外乎：

一、我们是宇宙的中心（换句话说，我们觉得自己的故事是唯一的、最重要的、最精彩的）；

二、我们超脱于整个宇宙（我们先看到自己，然后把其他一切万事万物都归入无用之物——狗、秋千、内布拉斯加州、低垂的云，当然，还有其他各国各地的人民）；

三、我们是永生不死的（确实有死亡这么件事，不过，那是发生在别人身上的事，跟我没关系）。

如今，我们不再相信这些——智识上，我们更加成熟了。然而，这些想法却深深地铭刻在我们心里，甚至成为我们的生存准则，这造成我们往往将自己的需求凌驾于他人之上。尽管我们内心真正渴望的是少几分自私，多关注当下正在发生的一切，拥有更开阔的心胸，更懂得关爱他人，等等。

我们想要更善良，因为我们都曾经体验过善良的心态，而且喜欢那种感觉。

所以次等重要的问题是：我们如何才能做到这一点？如何变得更善良，心胸更开阔？怎样才能少几分自私？怎样才能专注当下而不是抱着虚妄的幻想？等等。

是的。这是个好问题。

可惜的是，我只剩三分钟了，却要回答这么大的问题。

那就让我这么说好了：方法是有的。而且你们都已经知道了，因为在我们生命中，我们的善心有时候高昂，有时候低迷，你们也很清楚自己在何时会趋向前一种情况而远离后一种情况。

我们已观察到善良的程度可以变化，这会带来有趣的结果：既然善良是可以增减的，我们就可以推论到它可以被强化。也就是说，一定有实践之道能让我们时而丰沛时而枯竭的善良之心变得强大而坚定。

教育就是个好办法；全心全意投入艺术也是好办法；祈祷很好；冥想也可以；与亲密的朋友坦诚谈话；或者找到一个传统理念支撑自己。应该承认，在我们的人生开始之前就有无数聪明人士也提出过同样的问题，并为我们留下了宝贵的答案。

如果我们不懂得从过往的历史中找到这些智慧的声音，那就太可惜了，找不到满意的答案也是自食其果。就好像是我们罔顾物理学已有的发现，或者无视医学已经研发出的脑部手术成就，空想着自己重新找出这些奥秘。

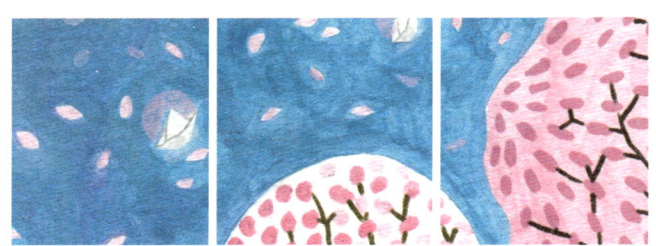

因为，与人为善这件事，其实远比我们所想的困难——这件事从我们面对美丽彩虹、可爱的小狗开始，慢慢扩展开来，几乎涵盖了世间万物的道理。

没错，就是这样。

有一个事实对我们是有利的：随着年龄增加，我们自然就会变得越来越善良。也许是单纯因为时间消耗掉了我们内心的自私。当我们的年龄逐渐增长，会不断体悟到自私自利是多么徒劳无益，并感受到这种行为并不符合人性的逻辑，真的。

当我们对他人表达爱意时，自我中心主义会得到纠正。当人生不如意时，当别人为我们辩护、对我们伸出援手时，我们会明白自己不是独立的存在，我们都想融入集体。当至亲至爱的人一个个离开这个世界，我们慢慢明白有朝一日，我们也会死去（但仍怀着一种心情：那一天离此刻的自己还很遥远）。大多数人都是越年老越善良。我觉得这再明确不过了。

雪城大学的伟大诗人海顿·卡鲁斯在他人生尽头所作的一首诗中说：他心里装着满满的爱，这一刻。

所以，让我预言，并且满心地期望，期望你们都会逐渐成熟起来，你们会更无私、更善良，自我中心会渐渐转化为对他人的爱。如果你有了小孩，摒除自我中心的进程会更快、更明显。那时，为了能让孩子受益，你会不惧任何困难。

你们的父母现在如此的自豪和快乐,原因就在于此。他们最大的梦想之一已经实现:你们学会了战胜困难而且已学有所成。成功让你们更加成熟。你们未来的生活会充满阳光,从这一刻开始,直到永远。

不过我还是要说,恭喜你们毕业。

年轻时,我们容易焦虑,因为我们总是怀疑自己把一件事做好的能力。这种情绪当然有道理。我们能否成功?我们能否过上有保障的生活?而你们——尤其是出生在这个时代的你们,可能已经发现了这个问题:人的野心周而复始,无穷无尽。如果你高中学业优秀,就一心想进好大学;如果你大学成绩非凡,接着就想找个好工作;如果你在职场业绩出色,就又想着……

有野心并不是坏事，在我们决心变得善良之际，也同样包含完成自身的使命。不论你是实业家、功成名就者，还是梦想家，为了成为最出色的自己，我们必须怀着雄心，把自己最好的一面活出来。

然而，功名是靠不住的。对于"成功"，人人各有定义。成功不但不容易，并且就算达到了，总是还想要更上一层（追求成功，犹如攀登一座不断增高的山）。而真正危险的是，你的生活完全被追求成功所占据，而最后你发现，人生巨大的疑问却始终没有解答。

回首我人生的过往，许多事情让我如坠云雾之中，同时让我放逐了心中的善良。我困在那些焦虑、恐惧、不安和野心之中，并错误地认为只要功成名就，就能解除这些恼人的情绪。我以为只要我做得更好——更成功、更有钱、更有名，内心的惶惶不安就会结束。

如今，我几乎可以确定，我就是从毕业那一天起，便困在这错误的想法中。这么多年来，我心中当然有善良之念，但我总是想，先让我过了这学期、让我拿到学位、让我写完这本书、让我赢得这个工作任务、让我买了房、让我养大孩子，然后，这些事情都做到了，我就会开始对他人和世界表现出善意。只是，事情没有结束的时候，只会一件又一件地来，循环无止境，永远没完没了。

所以，我快点说吧，演讲最后我的建议就是：既然人生是一个不断变得善良的过程，不妨调快这个进程，加快你变得善良的步伐，现在就开始行动。虽然我们被私欲所迷惑，身患自私自利这种顽疾，但这并非不治之症。

为你自己好，当个好病人，当个积极的病人，甚至过分积极都没关系，努力找寻让你能不自私的特效药，在你接下来的人生中，活力满满地追求康复。找出怎么能让自己更善良的方法，打开自己的心，让你自身中最能善待他人最慷慨无私的那部分展现出来——全力去做，仿佛人生其他一切都不重要，只有这点至关紧要。

因为,
其他一切
真的
都
不重要。

请什么都去尝试吧,那些让你野心勃勃想追逐的梦想——周游各地;大把赚钱;成为名人;创新改革;引领时代;陷入情网;你可以创造财富,但失去财富也无妨;或者在丛林里的河水中裸泳(建议提前检查对猴子粪便的免疫能力)——然而,尽管你竭尽全力让自己变善良,也难免会犯错。要把更多精力投入解决人生的大问题上,如果只是专注于无聊琐事,最后只会变得碌碌无为。

如果你愿为善良而战,你闪光的灵魂将挣脱个性缺点的束缚,像莎士比亚、甘地、特蕾莎修女的灵魂一样熠熠生辉。越过一切阻挠,到达神圣光明的地方。

信仰它的存在，了解它、认识它、滋养它，坚持不懈地与人分享它的果实。

然后到某一天，也许八十年后，当你们一百岁，我一百三十四岁时，当我们善良得都让人觉得受不了时，请给我写封信，讲讲你们的生活。我希望听到你们说：人生真美好。

我祝福大家幸福满满，愿大家好运常伴，并度过一个美好的夏天。

乔治·桑德斯写给年轻人的话

🍀 如果你感到困惑，不要担心。试着去永远保持这份困惑吧。一切皆有可能。永远保持开放的心态，走出自己的舒适区，然后走入更广阔的世界。一直这样做下去，直到你死去的那一天。阿门。

🍀 Don't be afraid to be confused. Try to remain permanently confused. Anything is possible. Stay open, forever, so open it hurts, and then open up some more, until the day you die, world without end, amen.

🌿 我不认为天底下有什么新鲜事。我们中的大多数人都和公元 1000 年的人一样过着日子：他们微笑着为生计奔波，在心底认为自己比那些笨蛋强得多；然而同时他们的另外一部分灵魂，也在为自己不如那些更聪明、更美丽的人们而感到悲伤。

🌿 I don't think much new ever happens. Most of us spend our days the same way people spent their days in the year 1000: walking around smiling, trying to earn enough to eat, while neurotically doing these little self-proofs in our head about how much better we are than these other slobs, while simultaneously, in another part of our brain, secretly feeling woefully inadequate to these smarter, more beautiful people.

🍀 在你的人生中可能会有一个这样的阶段：你决定止损，然后你还是吃亏了。接着你下定决心，这次真的要止损，然后你继续亏下去。这个倒霉的时期一直持续，你开始带着好奇看下去，想知道哪里才是自己人生的谷底。

🍀 There comes that phase in life when, tired of losing, you decide to stop losing, then continue losing. Then you decide to really stop losing, and continue losing. The losing goes on and on so long you begin to watch with curiosity, wondering how low you can go.

🍀 观万物,我们不过是"有错自远方来"的受害者……普世的人类法则——需求,对心爱之人的爱,恐惧,饥饿,间隙欣喜若狂,因无所惧、无所饥、无所痛而自然产生的仁慈善良——都是不变的、可预见的……何其强大,若我们知晓:个人的欲望竟然可以映射到陌生人身上。

🍀 In all things, we are the victims of The Misconception From Afar. ... The universal human laws—need, love for the beloved, fear, hunger, periodic exaltation, the kindness that rises up naturally in the absence of fear/hunger/pain—are constant, predictable. ... What a powerful thing to know: that one's own desires are mappable onto strangers.

🍀 先微笑,再说话。

🍀 ...smile first, then speak.

🍀 如果一个人没有持久的信念,没有持续的期待,没有对你愿意为之冲破所有暂时的禁忌、对心中正念的追求,那么对他来说什么才是真实的呢?

🍀 What is truth, if not an ongoing faith in, and continuing hope for, that which one feels and knows in one's heart to be right, all temporary and ephemeral contraindications notwithstanding?

🍀 我想让你读我的书，这本书对你非常重要。这并不是因为它可以提高你的文学素养，而是因为处于混乱状态的你真正需要它。

🍀 I want you to read my book and have it actually matter to you. Not to your constructed literary self. But to you. To the person who has issues and confusions.

🍀 我过去习惯向事物黑暗的一面靠拢——证明我是有料的；或者更急功近利一些，表明我对人性黑暗的认知程度。是的，这是真的，年龄增长了，我也更自信了。我现在认识到，人们可以把事情做得很好，或者人们可以比你我想象得更可爱。

🍀 My habit would have been to veer towards the dark— to prove I was something; edgy, or maybe to prove that I was cognisant of the dark side. Now, with age and confidence, I can say, yeah, that's true, but I am cognisant of the fact that people can do things well. And can be more loving than you expect.

一个人打破自己的习惯、懒惰以及有限的头脑，进而写出一些实际的东西，出版了，还对别人有用，这样的可能性微乎其微……即使成千上万的年轻人读后没什么收获，这过程对其而言仍然高尚。这个过程，是试图对其说点什么，是带领其逐渐体悟技巧问题、世界观问题以及自我意识问题——所有这些都是对性格的培养，而且，上帝保佑，我们所做的一切，可千万别仅仅是为了实实在在的职业生涯回报。我一次又一次看到，这种试图说点什么的过程给人以尊严，使人获得进步。

The chances of a person breaking through their own habits and sloth and limited mind to actually write something that gets out there and matters to people are slim... Even for those thousands of young people who don't get something out there, the process is still a noble one—the process of trying to say something, of working through craft issues and the worldview issues and the ego issues—all of this is character-building, and, God forbid, everything we do should have concrete career results. I've seen time and time again the way that the process of trying to say something dignifies and improves a person.

🍀 一生中，我们是很多个人的总和。

🍀 In our lives, we're many people.

如果你有消极倾向，可你却否认它，那么你这消极就翻倍了。如果你有消极倾向，你直视它，那么你就有可能转化它。

If you have a negative tendency and you deny it, then you've doubled it. If you have a negative tendency and you look at it, then the possibility exists that you can convert it.

🍀 学生水平这么高，你必须得真诚。这强迫你不断反观自己的进程，以免横生废话。

🍀 With this caliber of student, you have to be really honest. It keeps you looking at your own process, so you don't import any nonsense.

媒体和名人推荐

- （乔治·桑德斯是）极富想象力的作家，持续影响着一代青年作家，给当代美国小说带来了别具一格的幽默感、悲悯与文学风格。

  ——麦克阿瑟天才奖颁奖词

- 桑德斯的小说"阴暗得有趣"，探索了"处在普通与非凡压力下的人类自我"。

  ——首届弗里欧文学奖评委会主席　拉维尼娅·格林劳

- 乔治·桑德斯已经写出了你将在今年读到的最好的书。

  ——《纽约时报》

❦ （本文）如诗篇一样轻盈，但却深厚无与伦比。

——《纽约时报》

❦ 倘若真有"作家中的作家"，那就是桑德斯。

——**《纽约时报》前副主编 乔尔·洛弗尔**

❦ 谁的作品也没有乔治·桑德斯的作品有力量。

——**《纽约时报》特约书评人 角谷美智子**

🌸 没几个人能像桑德斯这样,深刻而又一针见血。

——**普利策奖、美国国家书评人奖得主　朱诺特·迪亚兹**

🌸 乔治·桑德斯——美国最令人兴奋的作家。

——**普利策奖提名人、小说家　大卫·福斯特·华莱士**

🌸 他可真是独一无二,他独树一格——但这一点众人皆知。

——**欧·亨利短篇小说奖、《爱尔兰时报》国际小说奖得主**

洛丽·摩尔

🌸 我认为他是在世的英语作家里面最会写短篇小说的。

——**怀丁作家奖、美国笔会非小说奖得主,畅销书作家**

玛丽·卡尔

要谈论桑德斯是如此之难,部分原因是作家群已达成共识,认为他不知何故反正不仅仅是一个作家,比作家要多一点什么……其写作像是出自某种圣人之笔。

——布克奖提名作家　约书亚·费里斯

他是过去二十年里我们文学史上的亮点之一。他的心灵如此宽宏大方、具有雅量,以至于你若在他周围做点小善事,你自己都觉得无地自容。

——斯坦福大学、雪城大学教授　图拜亚斯·沃尔夫

读者推荐

- Convocation speeches are not just for graduates, ever. They are for those who aspire, for those who hope one day to graduate to better life management, to a happier, more fulfilling existence, for those who still need advice. That probably includes all of us. Even George Saunders.

George Saunders is my kind of guru. He is funny, articulate, self-deprecating, smart, and unassuming. He doesn't pretend to be something he is not. He is a fiction writer who is flummoxed by humans, and yet is someone who has figured out a few things in his life.

毕业演讲不只是针对毕业生的,还是为那些心怀渴望的人,为那些盼望有朝一日能提升生活品质的人,为那些需要建议的人而服务的。这可能包括我们所有人。即使是乔治·桑德斯。

乔治·桑德斯是我的导师。他风趣、能言善辩、自谦、聪明、谦逊。他不会假装成另一副面孔。他是个虚构的作家,他被人类所迷惑,但在他身上却能发现一些东西。

——**雪城大学　艺术创作硕士**

- The speech is fantastic though, and this is a great "gift" book.

  这场演讲太棒了，这是一本了不起的礼物之书。

  ——普林斯顿大学　准毕业生

- The speech takes only about 15 minutes to read but each word carries weight and meaning. It is a speech to savor and re-read often. Although the speech is available online, I recommend buying the book. But in any event, the speech is absolutely worth reading.

  演讲只需要 15 分钟，但每个单词都有分量和意义。这是一场值得经常品味和重读的演讲。尽管这篇演讲在网上可以找到，我还是推荐你买这本书。无论如何，这个演讲绝对值得一读。

  ——刚走出校园的创业者

- This tiny book ... is sweet and funny and urges kindness above all else.

  这本小书……是甜蜜的，有趣的，并鼓励仁慈高于一切。

  ——一位家有毕业生的父亲

- He speaks truth in a world of "obfuscation".

  他在一个"困惑"的世界里说真话。

  ——美国哲学学会会员

- This book was short and sweet. Made me contemplate my relationships and how I treat others every day of my life.

  这本书短小而愉快。让我思考我的人际关系以及每天如何对待他人。

  ——互联网行业　上班族

🎈 So it is not just good for graduates, it is good for all adults.

这本书不仅对毕业生有益，对所有成年人都有启迪意义。

——加利福尼亚大学　社会学教授

🎈 Psychotherapists don't give gift books. Well, that's what I was taught in training, but 30 years of practice and a clear sense that this book would fill the bill, allowed me to bestow this thought in print to a client before my leave of absence from my practice. ... until the day she called to thank me for the gift book. ... Why else would anyone put an idea into book form? It's to change lives for the better...move us on.

心理治疗师不可以送书给别人当礼物——在职业培训中，他们就是这么教我的。但这本书满含30年人生阅历，有着清晰的智慧，所以我可以在告别职业生涯之前，将这本思想丰富的印刷书送给我的客户，总有一天，他们会为此打电话来感谢我。人们用书籍传递思想，还能有什么原因？都是为了提升生活……让我们取得进步。

——一位资深心理治疗师

🎈 A Commencement Speech that everyone should listen once!

每个人都应该听一次的毕业演讲!

——**美国好读网读者**

🎈 I love the tone and compassion in the lovely eloquent reminder of who we all can and should aspire to be.

我喜欢这个口气和慈悲心,可爱而意味深长地提醒我们都能够也应该成为什么样的人。

——**美国亚马逊读者**

附　写给毕业五年后的自己

英文版原文

Congratulations, by the way

Down through the ages, a traditional form has evolved for this type of speech, which is: Some old fart, his best years behind him, who over the course of his life, has made a series of dreadful mistakes (that would be me), gives heartfelt advice to a group of shining, energetic young people, with all of their best years ahead of them (that would be you).

And I intend to respect that tradition.

Now, one useful thing you can do with an old person, in addition to borrowing money from them, or asking them to do one of their old-time "dances", so you can watch, while laughing, is ask: "Looking back, what do you regret?" And they'll tell you. Sometimes, as you know, they'll tell you even if you haven't asked. Sometimes, even when you've specifically requested they not tell you, they'll tell you.

So: What do I regret? Being poor from time to time? Not really. Working terrible jobs, like "knuckle-puller in a slaughterhouse"? (And don't even ASK what that entails.) No. I don't regret that.

Skinny-dipping in a river in Sumatra, a little buzzed, and looking up and seeing like 300 monkeys sitting on a pipeline, pooping down into the river, the river in which I was swimming, with my mouth open, naked? And getting deathly ill afterwards, and staying sick for the next seven months? Not so much. Do I regret the occasional humiliation? Like once, playing hockey in front of a big crowd, including this girl I really liked, I somehow managed, while falling and emitting this weird whooping noise, to score on my own goalie, while also sending my stick flying into the crowd, nearly hitting that girl? No. I don't even regret that.

But here's something I do regret:

In seventh grade, this new kid joined our class. In the interest of confidentiality, her Convocation Speech name will be "ELLEN". ELLEN was small, shy. She wore these blue cat's-eye glasses that, at the time, only old ladies wore. When nervous, which was pretty much always, she had a habit of taking a strand of hair into her mouth and chewing on it.

So she came to our school and our neighborhood, and was mostly ignored, occasionally teased ("Your hair taste good?"— that sort of thing). I could see this hurt her. I still remember the way she'd look after such an insult: eyes cast down, a little gut-kicked, as if, having just been reminded of her place in things, she was trying, as much as possible, to disappear. After awhile she'd drift away, hair-strand still in her mouth. At home, I imagined, after school, her mother would say, you know: "How was your day, sweetie?" and she'd say, "Oh, fine." And her mother would say, "Making any friends?" and she'd go, "Sure, lots."

Sometimes I'd see her hanging around alone in her front yard, as if afraid to leave it.

And then—they moved. That was it. No tragedy, no big final hazing.

One day she was there, next day she wasn't.

End of story.

Now, why do I regret that? Why, forty-two years later, am I still thinking about it? Relative to most of the other kids, I was actually pretty nice to her. I never said an unkind word to her. In fact, I sometimes even (mildly) defended her.

But still. It bothers me.

So here's something I know to be true, although it's a little corny, and I don't quite know what to do with it:

What I regret most in my life are failures of kindness.

Those moments when another human being was there, in front of me, suffering, and I responded ... sensibly. Reservedly. Mildly.

Or, to look at it from the other end of the telescope: Who, in your life, do you remember most fondly, with the most undeniable feelings of warmth?

Those who were kindest to you, I bet.

It's a little facile, maybe, and certainly hard to implement, but I'd say, as a goal in life, you could do worse than: Try to be kinder.

Now, the million-dollar question: What's our problem—Why aren't we kinder?

Here's what I think:

Each of us is born with a series of built-in confusions that are probably somehow Darwinian. These are: (1) we're central to the universe (that is, our personal story is the main and most interesting story, the only story, really); (2) we're separate from the universe (there's us and then, out there, all that other junk—dogs and swing sets, and the State of Nebraska and low-hanging clouds and, you know, other people), and (3) we're permanent (death is real, O.K., sure—for you, but not for me).

Now, we don't really believe these things—intellectually we know better—but we believe them viscerally, and live by them, and they cause us to prioritize our own needs over the needs of others, even though what we really want, in our hearts, is to be less selfish, more aware of what's actually happening in the present moment, more open, and more loving.

We know we want to be these things because from time to time we have been these things—and liked it.

So, the second million-dollar question: How might we DO this? How might we become more loving, more open, less selfish, more present, less delusional, etc., etc?

Well, yes, good question.

Unfortunately, I only have three minutes left.

So let me just say this. There are ways. You already know that because, in your life, there have been High Kindness periods and Low Kindness periods, and you know what inclined you toward the former and away from the latter. It's an exciting idea: Since we have observed that kindness is variable, we might also sensibly conclude that it is improvable; that is, there must be approaches and practices that can actually increase our ambient level of kindness.

Education is good; immersing ourselves in a work of art: good; prayer is good; meditation's good; a frank talk with a dear friend; establishing ourselves in some kind of spiritual tradition—recognizing that there have been countless really smart people before us who have asked these same questions and left behind answers for us. It would be strange and self-defeating to fail to seek out these wise voices from the past—as self-defeating as it would be to attempt to rediscover the principles of physics from scratch or invent a new method of brain surgery without having learned the ones that already exist.

Because kindness, it turns out, is hard—it starts out all rainbows and puppy dogs, and expands to include ... well, everything.

One thing in our favor: some of this "becoming kinder" happens naturally, with age. It might be a simple matter of attrition: as we get older, we come to see how useless it is to be selfish—how illogical, really. We come to love other people and are thereby counter-instructed in our own centrality. We get our butts kicked by real life, and people come to our defense, and help us, and we learn that we're not separate, and don't want to be. We see people near and dear to us dropping away, and are gradually convinced that maybe we too will drop away (someday, a long time from now). Most people, as they age, become less selfish and more loving. I think this is true. The great Syracuse poet Hayden Carruth said, in a poem written near the end of his life, that he was "mostly Love, now".

And so, a prediction, and my heartfelt wish for you: as you get older, your self will diminish and you will grow in love. YOU will gradually be replaced by LOVE. If you have kids, that will be a huge moment in your process of self-diminishment. You really won't care what happens to YOU, as long as they benefit. That's one reason your parents are so proud and happy today. One of their fondest dreams has come true: you have accomplished something difficult and tangible that has enlarged you as a person and will make your life better, from here on in, forever.

Congratulations, by the way.

When young, we're anxious—understandably—to find out if we've got what it takes. Can we succeed? Can we build a viable life for ourselves? But you—in particular you, of this generation—may have noticed a certain cyclical quality to ambition. You do well in high-school, in hopes of getting into a good college, so you can do well in the good college, in the hopes of getting a good job, so you can do well in the good job so you can...

And this is actually O.K. If we're going to become kinder, that process has to include taking ourselves seriously—as doers, as accomplishers, as dreamers. We have to do that, to be our best selves.

Still, accomplishment is unreliable. "Succeeding", whatever that might mean to you, is hard, and the need to do so constantly renews itself (success is like a mountain that keeps growing ahead of you as you hike it), and there's the very real danger that "succeeding" will take up your whole life, while the big questions go untended.

I can look back and see that I've spent much of my life in a cloud of things that have tended to push "being kind" to the periphery. Things like: Anxiety. Fear. Insecurity. Ambition. The mistaken belief that enough accomplishment will rid me of all that anxiety, fear, insecurity, and ambition. The belief that if I can only accrue enough—enough accomplishment, money, fame—my neuroses will disappear. I've been in this fog certainly since, at least, my own graduation day. Over the years I've felt: Kindness, sure—but first let me finish this semester, this degree, this book; let me succeed at this job, and afford this house, and raise these kids, and then, finally, when all is accomplished, I'll get started on the kindness. Except it never all gets accomplished. It's a cycle that can go on ... well, forever.

So, quick, end-of-speech advice: Since, according to me, your life is going to be a gradual process of becoming kinder and more loving: Hurry up. Speed it along. Start right now. There's a confusion in each of us, a sickness, really: selfishness. But there's also a cure.

Be a good and proactive and even somewhat desperate patient on your own behalf—seek out the most efficacious anti-selfishness medicines, energetically, for the rest of your life. Find out what makes you kinder, what opens you up and brings out the most loving, generous, and unafraid version of you—and go after those things as if nothing else matters.

Because, actually, nothing else does.

Do all the other things, the ambitious things—travel, get rich, get famous, innovate, lead, fall in love, make and lose fortunes, swim naked in wild jungle rivers (after first having it tested for monkey poop)—but as you do, to the extent that you can, err in the direction of kindness. Do those things that incline you toward the big questions, and avoid the things that would reduce you and make you trivial. That luminous part of you that exists beyond personality—your soul, if you will—is as bright and shining as any that has ever been. Bright as Shakespeare's, bright as Gandhi's, bright as Mother Teresa's. Clear away everything that keeps you separate from this secret luminous place. Believe it exists, come to know it better, nurture it, share its fruits tirelessly.

And someday, in 80 years, when you're 100, and I'm 134, and we're both so kind and loving we're nearly unbearable, drop me a line, let me know how your life has been. I hope you will say: It has been so wonderful.

I wish you great happiness, all the luck in the world, and a beautiful summer.

图书再版编目（CIP）数据

人生最好的毕业礼 /（美）乔治·桑德斯著；徐之野译 . -- 海口：南海出版公司，2018.7
ISBN 978-7-5442-9254-2

Ⅰ.①人… Ⅱ.①乔…②徐… Ⅲ.①人生哲学–青年读物 Ⅳ.① B821-49

中国版本图书馆 CIP 数据核字（2018）第 053500 号

著作权合同登记号：30-2017-123

CONGRATULATIONS, BY THE WAY by George Saunders
Copyright ©2014 by George Saunders
Chinese (Simplified Characters) copyright ©2018
by ThinKingdom Media Group Ltd.
Published by arrangement with ICM Partners
through Bardon-Chinese Media Agency
ALL RIGHTS RESERVED

人生最好的毕业礼

〔美〕乔治·桑德斯 著
徐之野 译

出　　版	南海出版公司　（0898）66568511
	海口市海秀中路 51 号星华大厦五楼　邮编　570206
发　　行	新经典发行有限公司
	电话（010）68423599
经　　销	新华书店

责任编辑	李玉珍
策　　划	好读文化
内文插画	喜　久
封面设计	所以设计馆
版式设计	小　虫

印　　刷	北京利丰雅高长城印刷有限公司
开　　本	787 毫米 ×1092 毫米　1/32
印　　张	5.25
字　　数	80 千字
版　　次	2018 年 7 月第 1 版
印　　次	2018 年 7 月第 1 次印刷
书　　号	ISBN 978-7-5442-9254-2
定　　价	49.80 元

版权所有，未经书面许可，不得转载、复制、翻印，违者必究。